PUBLISHED BY RUSSELL KING
On CreateSpace, a division of Amazon.com

For information, contact:
russkingassoc@aol.com

ISBN 978-1540547460

December, 2016

Russell Widens a Road

What Do Civil Engineers Do, Anyway?

Civil engineers are men and women that enjoy construction.

They work in the cold and the heat.

They work in the rain and snow.

But they get the beautiful weather days, too.

They get muddy and dirty.

They work when the contractor works.

During the day or overnight.

They produce documents, measure quantities,

and inspect materials.

They record what was done.

This book is dedicated to those hardy souls that provide inspection during construction.

Foreword:

So, you want to be a civil engineer? Did you know that they design, inspect construction of, and help to maintain roads, bridges, canals, dams, and buildings? It is the second-oldest engineering discipline, after military engineering.

Sub-disciplines of the civil engineering curriculum include, but are not limited to:

- Architectural
- Coastal
- Construction
- Earthquake
- Environmental
- Forensic
- Geotechnical
- Materials
- Municipal or urban
- Offshore
- Structural
- Surveying
- Transportation
- Water resources
- Wastewater

Civil engineers work for Federal, State, and Local government; and they work in the private sector, from something as simple as an individual homeowner to much more complicated local, national, or international, companies.

The author is a retired civil engineer who spent over 40 years working on civil engineering projects – mostly roadway design and construction; but also, sewer and water installations; transportation engineering; and construction and land surveying. His work was done for Local and State governments as well as the private sector. He is a registered professional engineer with a degree in Aerospace Engineering from Iowa State University – *"Yes, as a matter of fact, it is Rocket Science!"*

So, you see, although you need an engineering degree, you might not need it to be in a specific curriculum. You can become registered as a professional in other fields as long as you have the education, experience, and the inherent knowledge and abilities that a qualified engineering program will provide.

You need the ability to think. To be a good engineer, you need to be able to picture in your mind what is behind the wall and then be able to draw it, modify it to make it do what you want it to do, and then to construct it. And, you need to be ethically honest about what you are doing and why it is important. If you messed up, you need to correct it, not blame someone or something else for it. Fix it, and move on.

If you want to be in charge, then you need to be a registered Professional Engineer. Only a licensed PE can be insured which is very important in the complicated projects that civil engineers work on. But maybe you'd be more comfortable as a non-professional – they also provide extensive guidance and support on construction engineering teams. You still get to widen the road or build the bridge, tunnel, or airport, and that is the real challenge. Maybe you'd be more comfortable outside of construction - in design, finance, or government?

But remember, a reasonable knowledge of construction will make you better in all other phases of civil engineering. So, be on a construction team during your early experiences to see how the plans you prepared were constructed. Mess around in the mud.

Experience the cold of a northern Midwestern winter to understand what the contractor's guys must do, see how errors or discrepancies on the plans need to be addressed in the field and how contractors do it, learn what problems occur with paying for the additional time required to make the corrections, understand how important it is for you to think about all possibilities during the design because they will occur in the field.

You choose what is most comfortable to you – but remember, all your actions are important for the public improvements that are the end product of the civil engineering process.

Russell King, PE

Introduction:

This book uses detailed photographs and brief descriptions to show how civil engineers were involved in a project to widen a pavement. It includes a little bit about some basic design concepts, but, mostly, it will show how construction proceeded, based on the design. It will show how they helped to control its construction, what they tested and measured, and what components were used to make it modern, safe, and structurally sound.

Guidance provided by civil engineers for road construction is similar for other infrastructure projects:
- Airports.
- Bridges
- Buildings
- Electric, cell, and telephone towers
- Petroleum and water wells and storage facilities
- Power plants
- Sewers and sewage treatment plants
- Tunnels
- Underground and surface mining
- Water distribution, treatment, and storage facilities

Civil engineers play vital roles in designing, permitting, financing, and constructing these projects, plus many more. They work on huge projects that cost millions of dollars or very small projects that cost only several thousand dollars. How large is "Large"? The design fee for a new atomic power plant is based on more than 500,000 man-hours – one engineer, working with no overtime, works 2,080 hours per year. That is considered a large project.

There is, literally, a whole world of infrastructure, all around the globe, that civil engineers are involved with.

Most roadways in America are owned and controlled by the government, whether that is Federal, State, Local, or a Tollway Authority. Other agencies that control roadways include Park Districts, Universities, and Airports. Most agencies have a separate department identified as the Department of Transportation (DOT) and they are all staffed with civil engineers.

Civil engineers study increased traffic on roadways caused either by general area wide population growth or specific property development. They use Level of Service (LOS), Average Daily Traffic (AADT), and Peak Hour Volume to determine how much traffic the roadway can handle, and they design changes to alleviate it.

Traffic is defined to be cars, trucks, motorcycles, and busses, along with pedestrians and bicyclists.

The most common methods to mitigate increased traffic include:
- Widen the pavement to create left turn, right turn, and/or additional through lanes.
- Change the traffic control from stop or yield signs to traffic control signals.
- Use roundabouts or other unique intersection designs.
- Interconnect a series of traffic signals so they operate as a system and progress traffic, rather than allowing the intersections to act individually and randomly. This will maximize the time available for the main traffic movement.

Typical DOT Map of Average Daily Traffic:

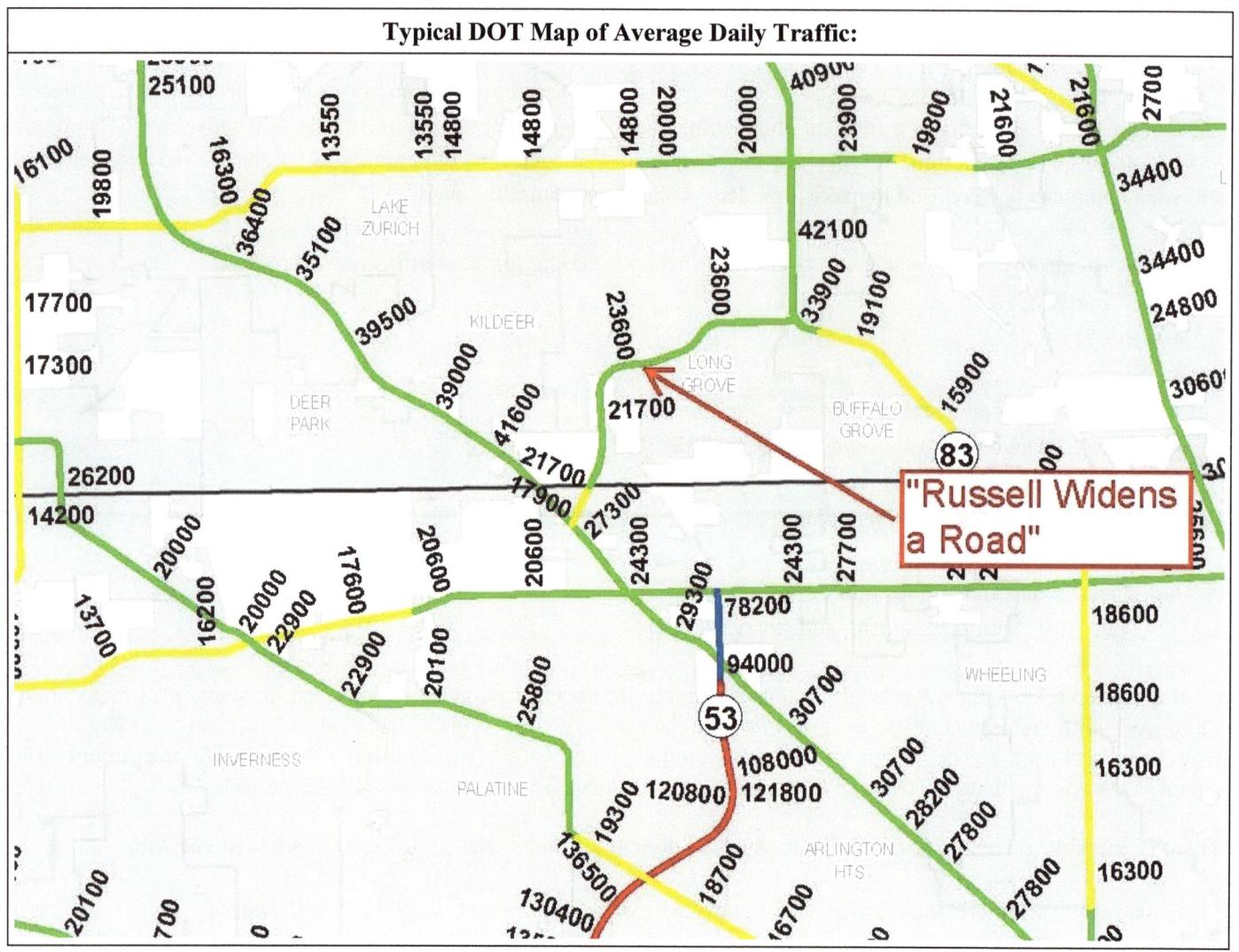

Construction Management Team: Civil engineers on a construction site form a Construction Management (CM) Team to provide the following services:

- Inspect and document the work completed by the contractor.
- Test the materials used for construction and document the results.
- Act as liaison between the contractor, the designer, and the DOT for questions and issues raised during construction.
- Coordinate day-to-day activities between construction, maintenance of traffic, ROW acquisition, erosion control, and utility relocation.
- Review and recommend approval of changes to the plans due to unforeseen issues.
- Keep a record of all Plan changes and prepare an "As Built" set of plans to identify the changes.

For purposes of this book, the term "Civil Engineer" will be used for civil engineers educated, trained, and licensed as Professional Engineers; Construction Technologists with construction technology training and education; and Technicians or Inspectors, either with or without formal engineering education but with experience in the civil engineering field.

For the Project, the CM Team consisted of one person, that person was a PE, and, since it was a small project, provided complete inspection services. On larger projects the Team might be composed of some, or all, of the following civil engineers:

Resident Engineer (known as the "RE") -- must be a PE:
- In day-to-day charge of all construction engineering and the CM Team.
- Assign persons to inspect, measure, and document construction, in coordination with the contractor.
- Coordinate with the owner's project manager.
- Act as the spokesman for the owner with the contractor, utility companies, and outside agencies.
- Coordinate lane closures between the contractor and the owner.
- Review and approve pay estimates.
- Review and approve shop drawings, methods of construction, questions about construction, traffic control and protection, and all related submittals.

Documentation Specialist/Assistant Resident Engineer -- should be a PE:
- Review and enter reports, information, and data from the CM Team into the Project Management Program.
- Check calculations prepared by the CM Team for accuracy, make corrections as required, and attach appropriate supporting data.
- Keep a record of the quantity of materials installed and inspected.
- Prepare pay estimates, extra work orders, and change orders.
- Keep meeting notes, prepare minutes, and send them to participants.
- Support the RE with contractor submittals, projected work schedule, methods of construction, and construction issues.
- Keep a daily record of weather, construction equipment and personnel, and work accomplished.

Material Coordinator-- could be a PE:
- Review and approve material mix designs for concrete, bituminous concrete, and fill materials.
- Work with the owner to coordinate materials from all phases and other projects on the system.
- Work with the Quality Assurance Inspector (QA) to periodically inspect material as defined in the contract documents.
- Work with the Quality Control Inspector (QC) to verify that all testing is being completed, as defined in the contract documents, and that all reports are completed and provided to the RE.

Technicians and Inspectors, usually will not be PE's, but they will have training in Construction Technology:
- Measure and document work completed by the contractor he/she is assigned to work with.
- Record his/her observations, measurements, and data in a field book and transfer the appropriate information to an Inspector's Daily Report (IDR).
- The Quality Assurance (QA) Inspector performs field tests for compaction, concrete, bituminous, bolt tightening, and stud shear connections and documents the results.
- Call to report lane closures, inspect them, and call in when they are reopened.
- Inspect and document the condition of traffic control, erosion control, and maintenance of traffic.
- Receive, initial, and tabulate load tickets.

Surveyors, licensed as Professional Land Surveyor (PLS):
- Use survey equipment to verify the final position of major elements including Temporary Retaining Walls, Piles, Permanent Retaining Walls, Piers and Abutments, Beam positions, Fillets, and Finished Bridge Deck Elevation.
- Use survey equipment to continually check the furnished embankment elevation as it approached the limits.
- Use survey equipment and computer programs to measure and compute the amount of topsoil stripped and the amount of embankment furnished and compacted.

Professional Engineer: A Professional Engineer (PE) must have the following education and training:

- Graduate with at least a Bachelor of Science (BS) from an Accreditation Board for Engineering and Technology (ABET) accredited engineering school.
- Take and pass a standard Fundamentals of Engineering written exam (FE) -- usually taken immediately upon graduation.
- Accumulate a minimum of four years training under the direction of Professional Engineers.
- Complete and pass a written Principles and Practice in Engineering (PE) examination.

Each state has the authority and requirement to license PEs. The PE license must be renewed every two years. During the two years, the Engineer must certify that they worked on engineering projects, that they accrued professional development hours in their specialized field, and that they paid the appropriate fee.

The requirements are established and defined by states. Some states issue generic professional engineering licenses while others issue licenses for specific disciplines of engineering, such as Civil, Structural, Mechanical, Nuclear, Electrical, or Chemical. However, in all cases, engineers are ethically required to limit their practice to their area of competency. In some states, licensed civil engineers may also perform land surveys.

Project Management Programs: Many project owners elect to use a project management computer program to manage documents that are generated by large projects. The programs provide a structure so all persons who have an interest in the project, and are authorized to be involved, can have access to the documents. It also keeps track of the people who are required to respond to the document. These management programs require a CM Team member to input data daily so that the program is up-to-date and can manage the information while helping to keep important information moving.

QA and QC Inspectors: The contractor is ultimately responsible for, and must test and accept, all materials used on the project. This requirement is well defined in the contract documents and applies to both small and large projects.

On large projects, the contractor furnishes a civil engineer to act as an inspector called the Quality Control Inspector (QC). This person is responsible to perform the tests, prepare the reports, and certify that the materials meet the requirements. A second inspector, provided by the CM Team and called the Quality Assurance Inspector (QA), periodically tests the same items but at a reduced rate, again, as identified in the contract documents. This cross-checking assures that the materials meet the requirements of the contract documents.

On the Project, the contractor furnished the QC person, and the CM Team furnished the Resident Engineer who verified that the test(s) were completed, and the results met the requirements.

"Yield": A yield check is a calculation done by the civil engineer to verify that enough material was delivered for each constructed item. So, for instance, for curb and gutter, the concrete truck drivers gave load tickets to the civil engineer. At the end of the day they tabulated them to show how much material was delivered. They also measured how much curb and gutter was installed, calculated the volume, and then verified that enough material was brought. For the following example of Type B-6.24 curb and gutter, the contractor had to deliver at least 41.4 cubic yards.

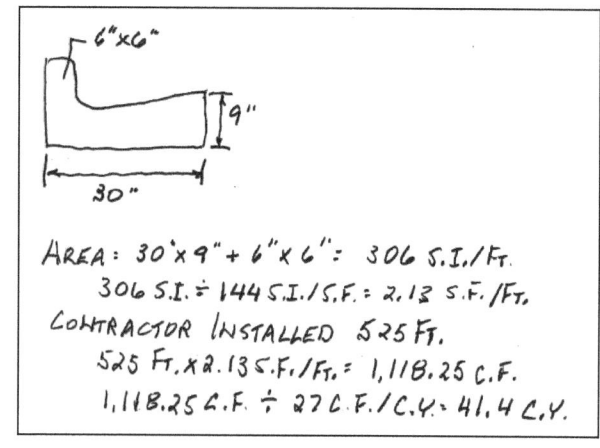

Surveyor: Surveying is a specialized study in the Civil Engineering curriculum. It is used to obtain and plot topography, ownership, and distances and angles between points on the earth. Surveyors are state licensed as a Land Surveyor in Training (LST) after passing an eight (8) hour test and then a Registered Land Surveyor (RLS) after four (4) more years of supervised work under a registered land surveyor and after passing another eight (8) hour test.

Inspector's Daily Report (IDR): The IDR is a document created to consistently record certain information obtained in the field including pay items, weather, equipment, measurements, sketches, recorded information, and other observations made by the civil engineer. Additional information, such as copies of the field book pages, plan sheet details, calculations, and other supporting data are attached to the IDR, as supporting documentation, checked by the Documentation Specialist, and approved by the RE.

Each CM Team member completes an IDR every day that he/she provides inspection. Prior to the proliferation of computers and other electronic equipment, it was completed on paper by hand. As computers become more and more mobile and field friendly, they will be used directly in the field to take the place of field books and hand written documents. The important thing is that the CM Team member must properly document what was done, how it was inspected, and that it met the contract documents.

A typical Inspectors Daily Report (IDR):

Typical Field Book page attached to the IDR:

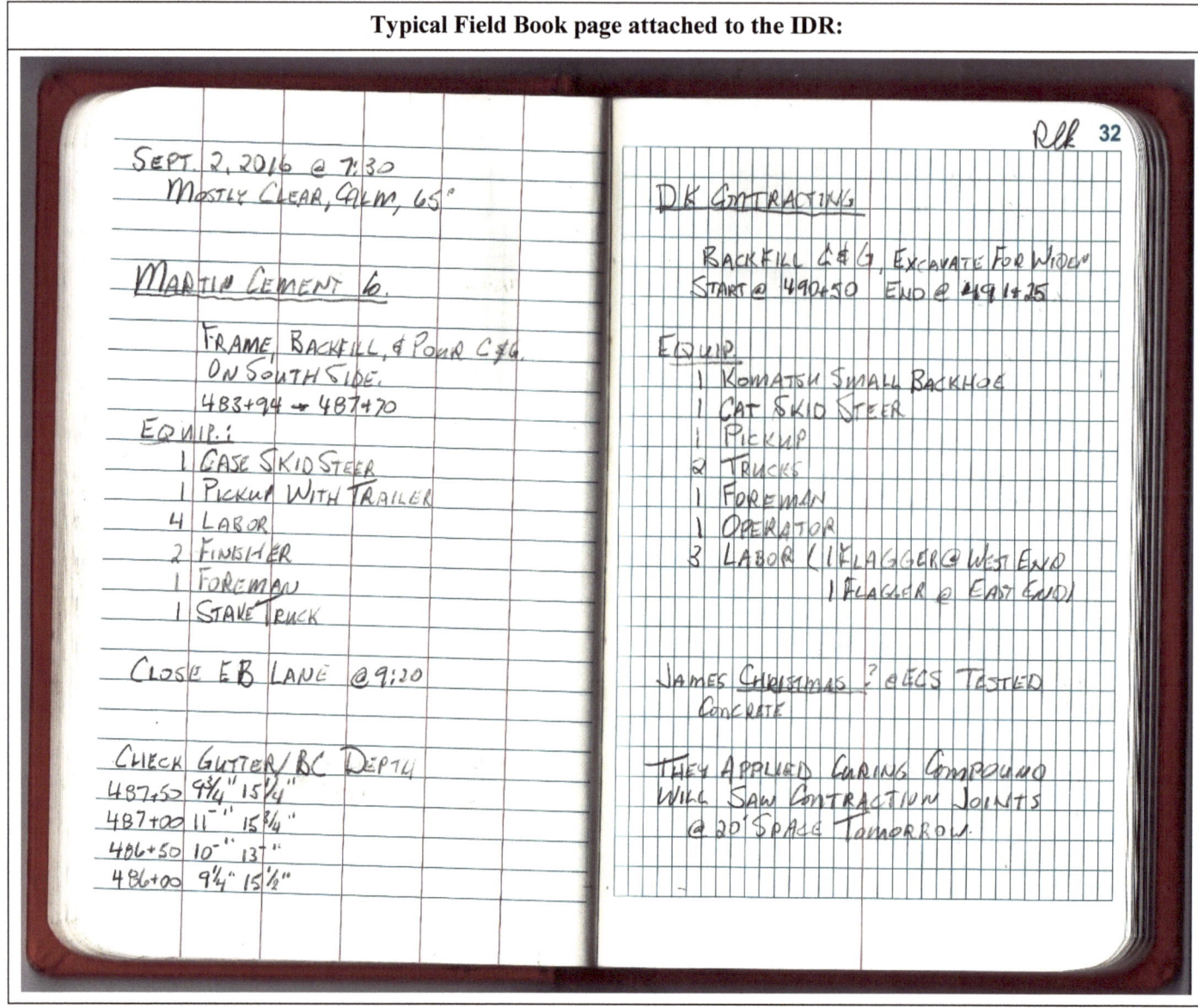

Paying the contractor: "Russell Widens a Road" was a private project. As such, the contractor bid on it, gave a "lump sum" price to the developer, and the developer paid it. The CM Team was not concerned about how much work the contractor performed to complete it. Any deviations and/or extra work items had to be negotiated directly between the developer and the contractor. The CM Team only had to make sure that the work was completed as required.

If this had been a DOT sponsored project, the designer would have included a detailed list of quantities in the contract documents, the DOT would have used that list to prepare a budget and the contractor would supply unit prices for each item on the list for the bid. The civil engineer would measure and verify the installed quantity of each item, multiply it by the unit price, and pay the contractor that amount. If there was an increase between the plan quantity and the installed quantity, the contractor would have to justify it, submit it to, and have it reviewed by, the CM Team, and then approved by the DOT. The contractor would not be paid for excess material until the review and approval process was finished.

Contract Documents: Contract Documents are written documents that define the roles, responsibilities, and "Work" under the contract. They are legally binding on both the "Owner" and the "Contractor". Many documents are combined to create the "Contract Documents" - Plans, Details, Special Provisions, Standard Specifications, Shop Drawings, and Contracts are a few of the more important ones. Civil engineers are heavily involved with all phases of contract document preparation and work closely with the owner, as well as their financial, legal, and planning professionals – many of whom are also civil engineers.

The Project:

"Russell Widens a Road" is a project to widen a pavement owned and controlled by the DOT. A developer initiated and paid for changes to it, with approval from the DOT, because the developer wanted a new access to a new senior care facility being constructed on an adjoining 9.23-acre parcel. The facility included several hundred small apartments, a health and community center, offices, and long term care areas, surrounded by parking and storm water management facilities.

A very simplified plan shows the work that must be done for the Project:

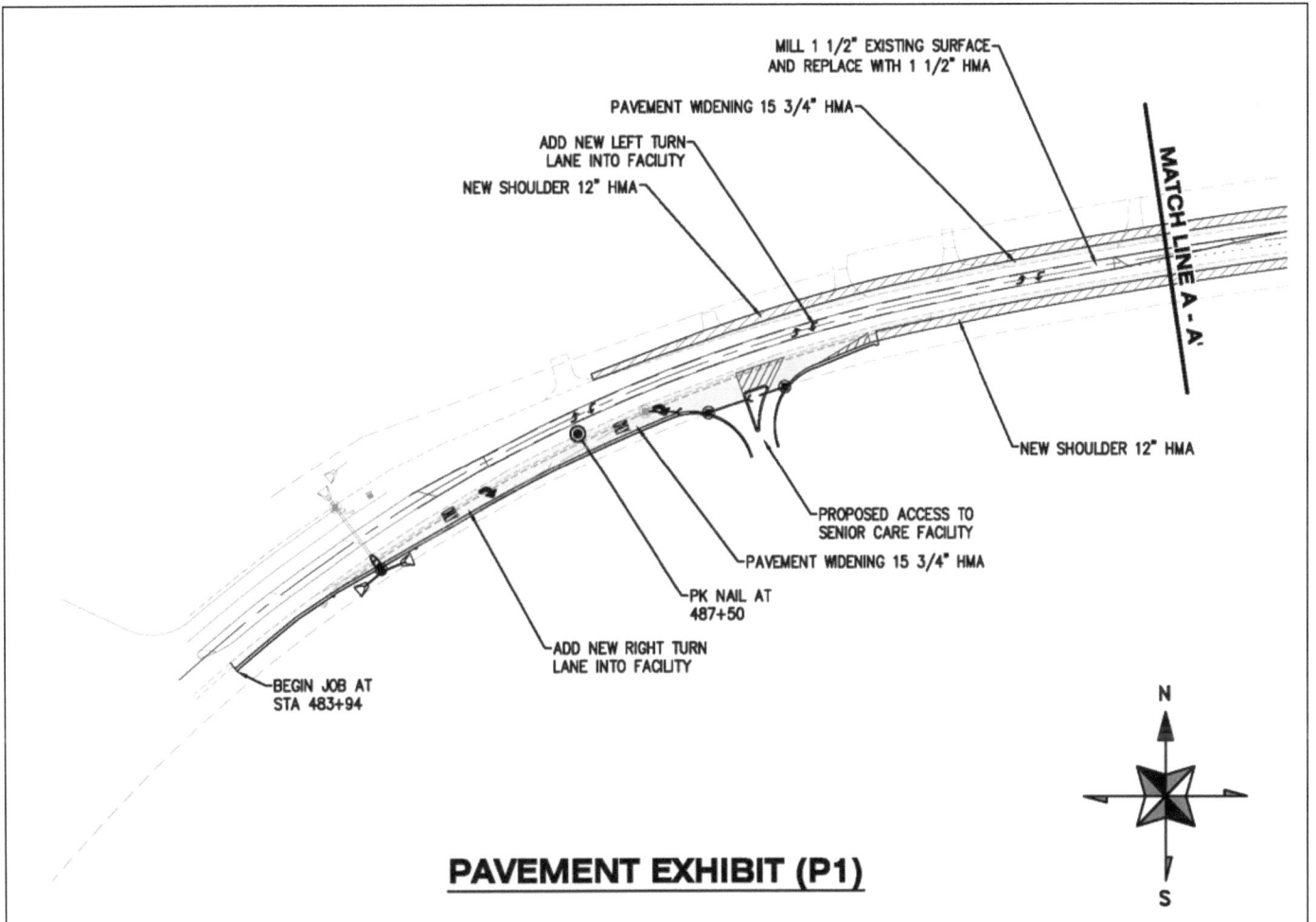

PAVEMENT EXHIBIT (P1)

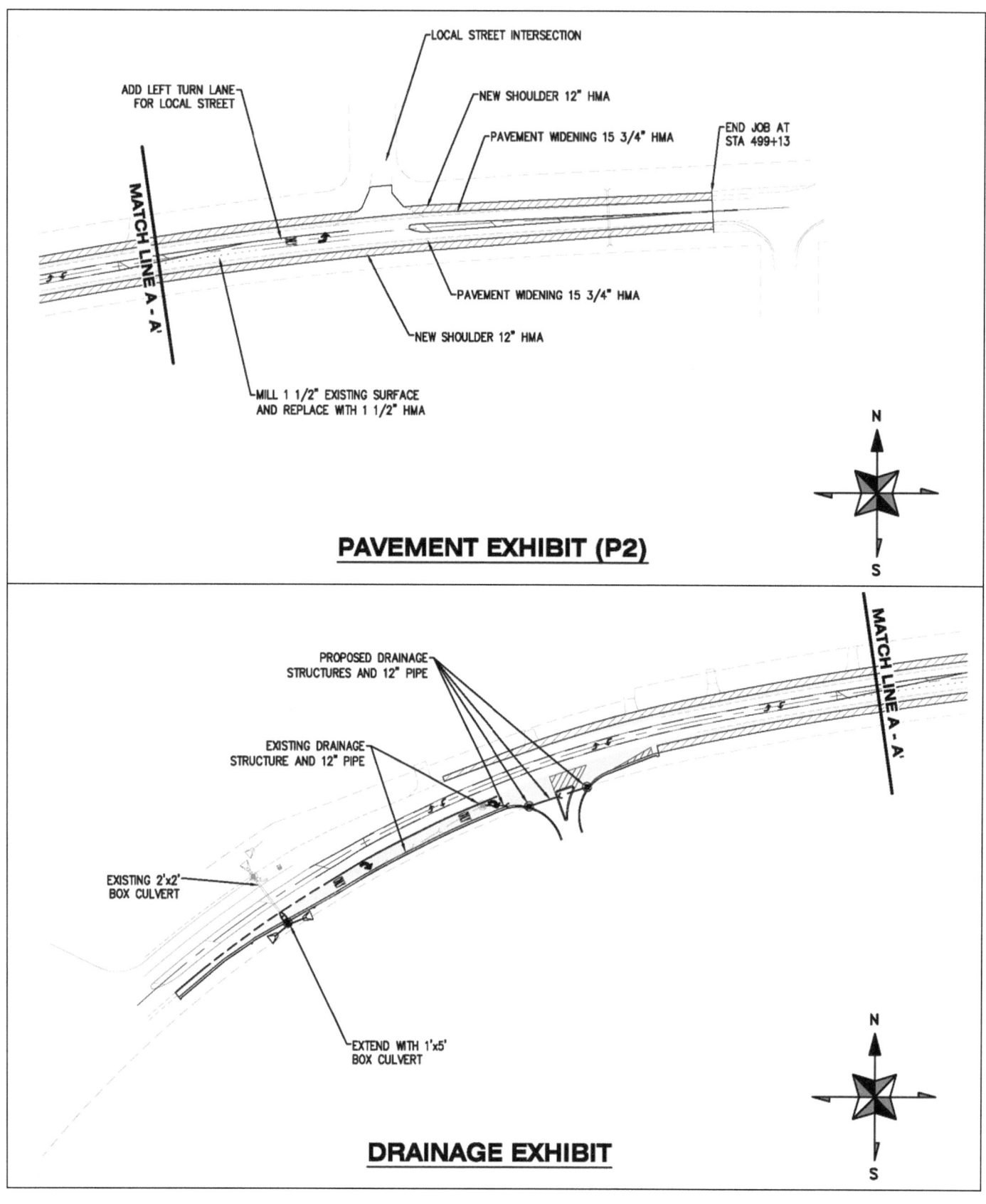

PAVEMENT EXHIBIT (P2)

DRAINAGE EXHIBIT

The Project increased traffic on the DOT Road, so they required the developer to retain a civil engineer to study it and see how much traffic the new facility would generate, how much traffic was on the existing highway, and how the additional traffic would impact the LOS. Once the civil engineer determined that the Project decreased the LOS, they worked with the DOT to recommend the addition of an eastbound right turn lane and a westbound left turn lane at the proposed driveway into the facility. The DOT also required an eastbound left turn lane at the next local street intersection.

The civil engineer designed the roadway and wrote the specifications. The civil engineer then created a CM Team to inspect construction to verify that it met the contract documents.

1,520 feet of existing pavement was widened to include the addition of left and right turn lanes and a shoulder on each side. The existing pavement surface was milled off and replaced with new surface to give a completely new pavement surface. Curb and gutter was added along part of the south side and storm sewer, ditches, and pavement marking were constructed. A new driveway was constructed to serve the facility.

The Project would be considered simple, especially since there were no traffic signals, and very typical of many pavement widenings throughout America. Even so, the civil engineer had to provide many of the same services required for large projects stretching thousands of feet or even several miles. Some of the major features in the basic design included:

- Obtained and plotted the existing topography.
- Selected a mix design for the widening consisting of 12 inches of "Hot Mix Asphalt" (HMA), 2 1/4 inches of Binder Course, and 1 1/2 inches of Surface Course, for a total of 15 ¾ inches of pavement.
- Selected 12 inches of HMA for the new shoulders.
- Milled and resurfaced the existing pavement.
- Designed drainage structures for proper roadway drainage.
- Generated cross sections to show how much cutting or filling was required to meet the proposed grade.
- Prepared special provisions and details.
- Submitted the plans to the DOT for review, approval, and the permit.

The two most common stabilized pavements in America are bituminous asphalt concrete and Portland cement concrete. HMA was the bituminous asphalt concrete selected for the Project. It is a typical bituminous concrete consisting of various sized aggregates (up to about 1 ½ inch size) bound together with liquid asphalt, laid in lifts (generally no more than 4"), and rolled to the proper compacted density. There are many different mix designs using various combinations of aggregates and liquid asphalt and all of them are created in a plant at over 300 degrees. The QC was responsible to provide the proper mix design to meet the contract documents.

Maintenance of Traffic: The Project roadway improvements were constructed on the existing two lane pavement with a very heavy AADT of 23,600. To control the traffic during construction, the contractor provided traffic control, including closing a single lane during working hours, with approved devices including barricades, cones, flaggers, or temporary barrier walls. Lights were attached to the devices for nighttime operation.

The civil engineer inspected the barricades daily to verify they were in good condition and properly spaced. Night inspections were held regularly to verify the condition of lights and reflectors.

Barricades, with lights, provide proper traffic control:

Typical Traffic Control for nighttime control:

Paving required one lane to be closed with flaggers and barricades:

Flaggers for Traffic Control: Flag persons were used to coordinate construction equipment with the motoring traffic. Properly placed and used, the flagger provides a very important advanced notice to reduce conflicts with vehicles using the pavement and to prevent crashes. The flaggers used portable paddle signs and flags to alert motorists to the upcoming construction activities.

Flagging involves knowing and understanding construction operations, vehicle movements, and potential conflicts between the construction and the vehicles. Flagger Certification is available from the American Traffic Safety Services Association (ATSSA), the National Safety Council (NSC), and the Union Laborers' & Contractors' Joint Apprenticeship Program at the state level. These certifications are available at many community colleges, through many labor unions, and other technical training agencies.

Flagger properly controlling traffic:

Stabilized Construction Entrance: The Project included a properly designed construction entrance to provide access to the construction site from the DOT highway.   The driveway was excavated, underlain with filter fabric, and constructed with crushed aggregate, properly placed and compacted. The entrance was maintained during construction and changed to the permanent entrance.

In addition to providing access, the stabilized construction entrance provided a staging area where mud and debris was removed from tires. The contractor was responsible to prevent mud and debris being deposited beyond the construction site and onto the DOT Highway. During muddy periods, the contractor even washed the tires of vehicles leaving the site so they would not track mud and/or debris onto the DOR pavement. Scrapers and sweepers were continuously used to keep the DOT highway surface clean.

2-inch crushed stone cleaned the tires before entering the DOT highway:

Washing tires before entering DOT highway:

Existing Utilities: There were both aerial and underground utilities within the DOT right of way (ROW) that conflicted with the Project and had to be removed by the utility owners. They either used their on-staff civil engineers or they hired other civil engineers to prepare plans to design the relocation and then hired their own contractors to do the relocation. For the Project, the utility owners took full responsibility and contracted the work themselves. The only work the CM Team did was to coordinate with them and verify that they were properly relocated.

Fiber Optic cable that had to be lowered:

Fiber Optic cable was lowered by trenching and lowering it:

Underground telephone lines that had to be relocated:

The CM Team can become involved with complicated conflicts to review and approve plans for relocation. This large diameter high pressure gas main had to be relocated on a larger project.

Erosion Control: The Project design engineer prepared a Storm Water Pollution Prevention Plan (SWPPP) which provided instructions and details to minimize erosion and non-point source pollution. It was submitted to, and received approval from, other civil engineers at the environmental protection agency. The civil engineer used the Plan to periodically inspect the erosion control features and prepared reports to show that the features were properly installed, maintained, and provided the necessary control.

Typical erosion control features include silt fence, ditch checks, and temporary and permanent seeding, covered with erosion control blanket.

What is Erosion? This is a natural response to wind and/or water passing across dirt from which the ground cover has been removed. When a road is under construction, the exposed dirt used for the road base, ditch section, or stockpiles can be blown or washed away.

Silt Fence to prevent sediment from leaving site:

Ditch Checks are another form of erosion control:

Erosion Control under snow:

Erosion Control – Silt Fence with wire support:

Excavation: The contractor saw cut, removed, and excavated the existing bituminous shoulder and enough material to construct the proposed pavement and shoulder widening.

The surveyor installed nails and stationing on the existing pavement surface as control points so the contractor could properly measure the depth and width required for the widening and shoulder.

The material that remained after removing the existing shoulder was called the "Subgrade". The civil engineer tested it to verify that it met the specifications and the widening and shoulder were constructed on it. See "Testing Subgrade".

When the subgrade material was not suitable, the civil engineer defined the limits of unsuitable material and worked with the contractor to undercut, reach suitable subgrade, properly grade and compact it, and then backfill with suitable material.

The civil engineer checked the yield - the actual undercut area times the thickness at the unit weight of the backfill.

The existing pavement was saw cut before it was removed:

The existing bituminous pavement was removed, loaded into trucks, and hauled off-site:

Suitable subgrade after pavement was removed:

Unsuitable subgrade was undercut 9-inches:

Undercut area was properly backfilled with crushed aggregate and compacted:

The civil engineer checked the final elevation of the subgrade:

Furnished Embankment: "Russell Widens a Road" did not require any fill. However, many projects require the contractor to furnish, place, and compact fill material up to several feet thick to meet proposed grades.

Suitable embankment material for fill includes crushed concrete, bituminous grindings, crushed aggregate, and embankment material. The material gradation must meet the requirements of the contract documents.

The following tests would be performed by both the QA and QC Inspectors:
- Verify that the type of material and gradations met the contract documents.
- Verify that the lifts were placed no more than 8 inches thick.
- Verify that the material was compacted to 95% Proctor (See Testing Compaction, below).
- Accept and sign for load tickets to verify that the proper material was delivered.
- Measure and pay for the total quantity of fill material brought and placed on site.

Beginning to place, grade, and compact fill.
Engineers check the density of the compacted material:

An 18 wheel semi-trailer, rear dump, can haul about 14 cu. yd. A dozer spread and a roller compacted it:

Cross Sections: When fill or excavation is required, cross sections are prepared by the design engineer and included in the plans. A mathematical formula called "average end area" is used to calculate how much fill or excavation is required. Both the engineer and the contractor use the cross sections to estimate the quantity that must be provided or excavated.

Cross sections also show the approved, finished, side slopes and how much topsoil is required.

When construction is completed, the surveyor will obtain the final elevations, plot them, and calculate the final quantity. The CM Team will then recommend paying the contractor this amount.

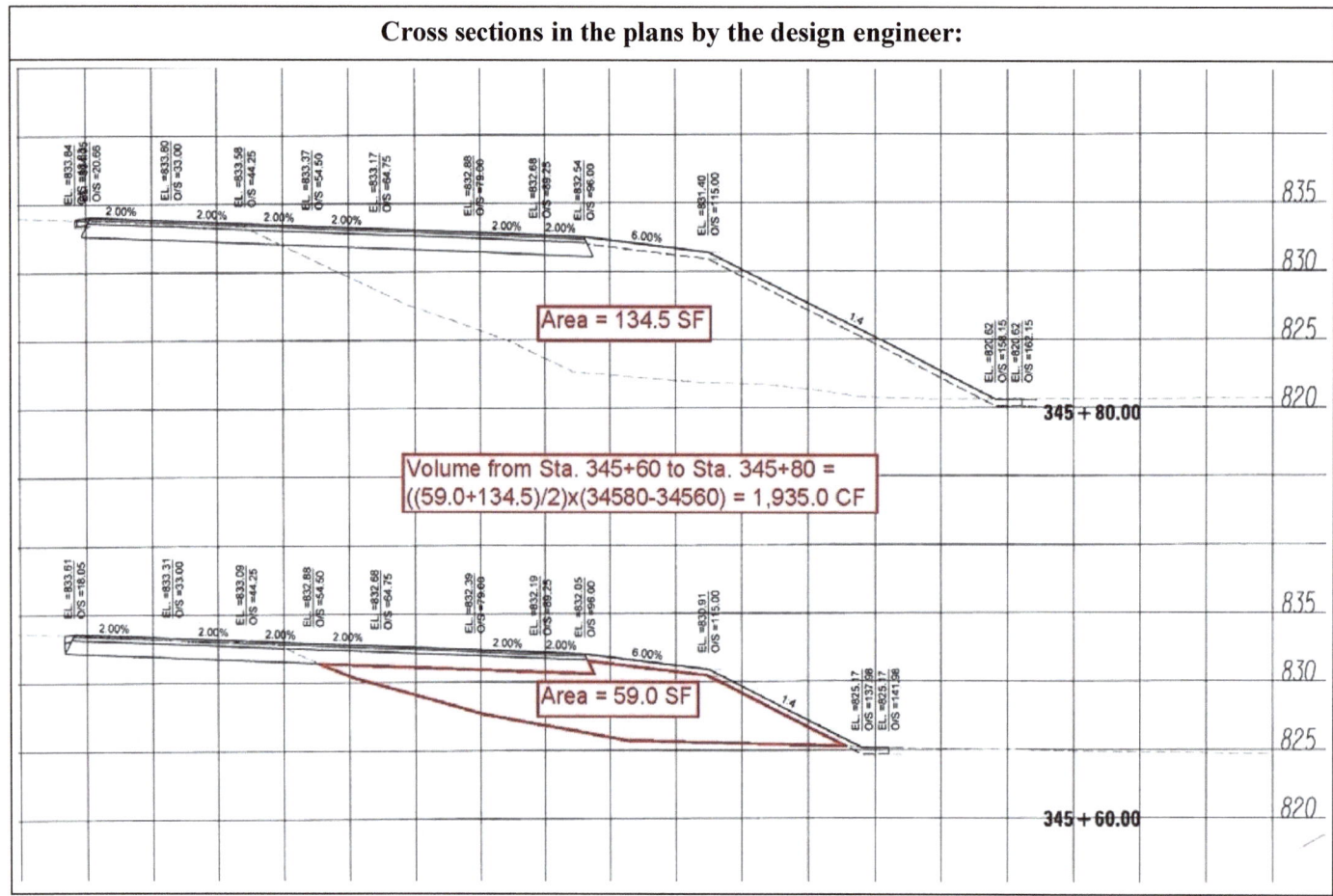

Testing Subgrade: After the subgrade was graded and compacted, the civil engineer tested it for density by using a Penetrometer, which is a probe to identify the in-place density of fine grained and granular materials or for weakly cemented materials. It will not work on course granular materials.

The probes are either pocket size or stand up length and they all have an attached meter to record the density. The probe is pushed into the material to create the "Test Force" and the meter shows how much resistance the material provides. The resistance is then converted to an approximate density. They do not measure moisture content.

Using the handheld (pocket size) Penetrometer. The Inspector applied the force:

Using a Static Cone Penetrometer. The Inspector applied the force by pushing into ground:

Using a Dynamic Cone Penetrometer. A dropped, movable weight, applies the force:

The only difference between Static Cone Penetrometer (SCP) and the Dynamic Cone Penetrometer (DCP) is how the testing force is applied:

- The force on the SCP is applied by the inspector with his weight and the resistance force is shown on the scale.
- The force on the DCP is applied with a 17.6-pound movable weight, raised and dropped 22.6 inches. The DCP records the density more accurately than the SCP by counting the number of blows required to achieve a calculated depth of penetration.

A Nuclear Density Gauge can also test the density and moisture of both granular and solid materials but was not used for the Project:

- A probe would be inserted into a hole approximately half way through the material being tested.
- A gauge emits a directed beam of atomic particles and a sensor counts the received particles reflected from the surrounding materials.
- The device converts the received bounce-back particle count into material density and moisture content.
- The density would be compared with, and recorded as, a percentage of Proctor. The contract documents would define the required density and moisture content.
- The Nuclear Density Gauge is the most accurate method currently available.

What is Proctor? Proctor is a laboratory method of experimentally determining the optimal moisture content at which a given material will become most dense and achieve its maximum dry density. It must be calculated for each type of material, and must be checked often for the same type of material because the material varies and will have different characteristics. The Proctor number is computed by a civil engineer in a laboratory situation and then used by the QA and QC Inspectors in the field.

Pavement and ROW Drainage:

- Storms cause rain to fall on the pavement and within the drainage area of it. *Civil engineers study storms, create mathematical models about how the rain falls, and use them to predict how much rain will fall.*

- A certain percentage of the rainfall will run off when it falls on the ground and pavement. *Civil engineers study the ground characteristics and use formulas to calculate how much runoff will be generated, based on the characteristics.*

- The runoff is collected by, and flows in, gutters and/or ditches constructed with the pavement. *Civil engineers use formulas to design proper longitudinal and transverse slopes for both the ditches and pavement so they can handle the runoff.*

- Drainage structures, with open grates, collect the runoff before it gets deep enough to encroach into the driving lane. *Civil engineers use formulas based on the pavement slope, the drainage area, and the grate capacity to calculate how far apart to put the drainage structures.*

- Storm sewers take the water from the drainage structures to discharge points. *Civil engineers use formulas based on the pipe slope and roughness to calculate how large the pipe needs to be to handle the water.*

Before the advent of computers, civil engineers solved the formulas by hand. Since they are complicated and time-consuming, they simplified and converted them to Nomographs which Wikipedia defines as ". . . *two-dimensional, graphical calculating diagrams, designed to allow the approximate graphical computation of a mathematical function.*"

The following typical Nomographs were used to determine:

- How much water a circular culvert will handle based on the diameter, length, and head.

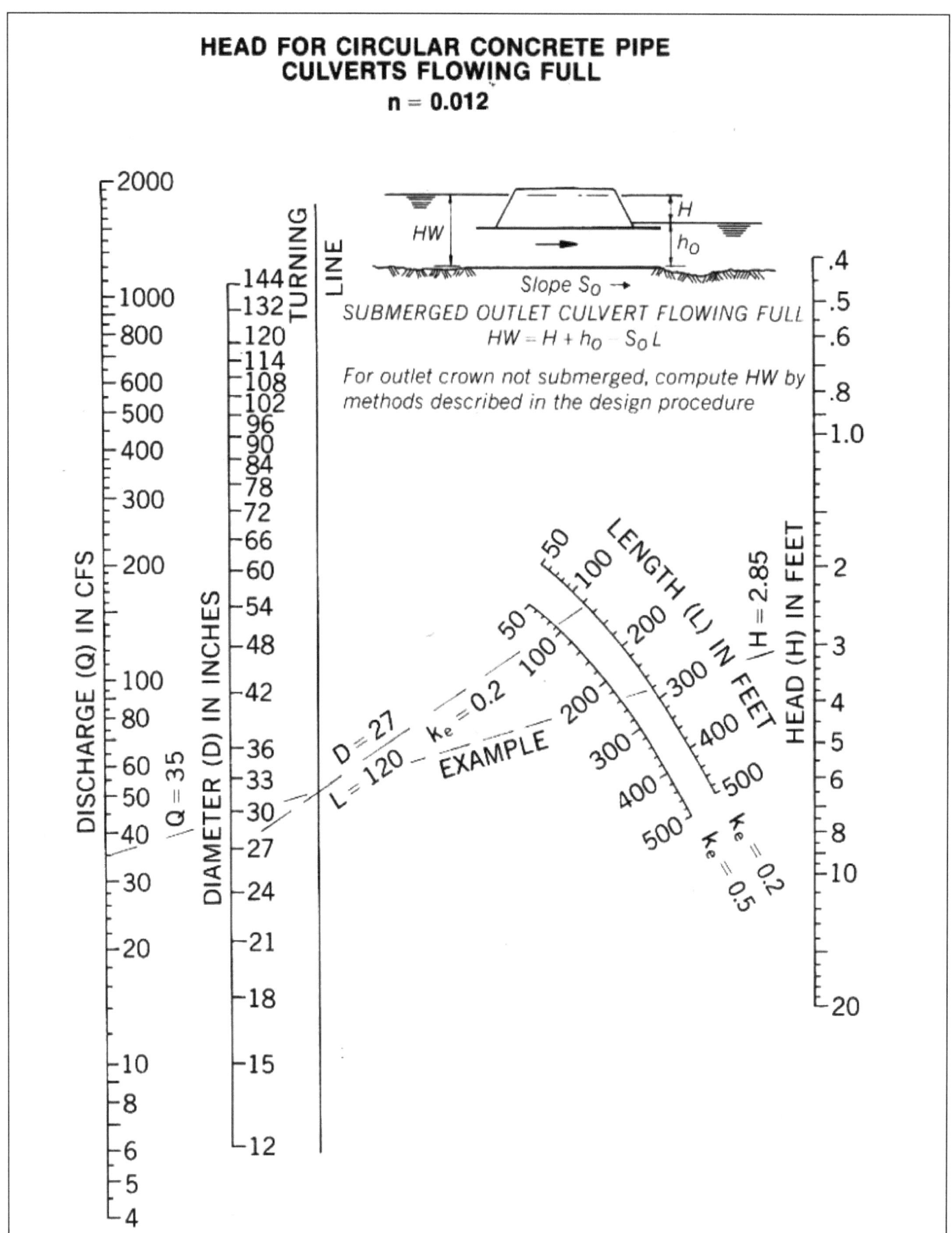

- How much water a circular storm sewer pipe can handle based on the roughness, slope, and diameter.

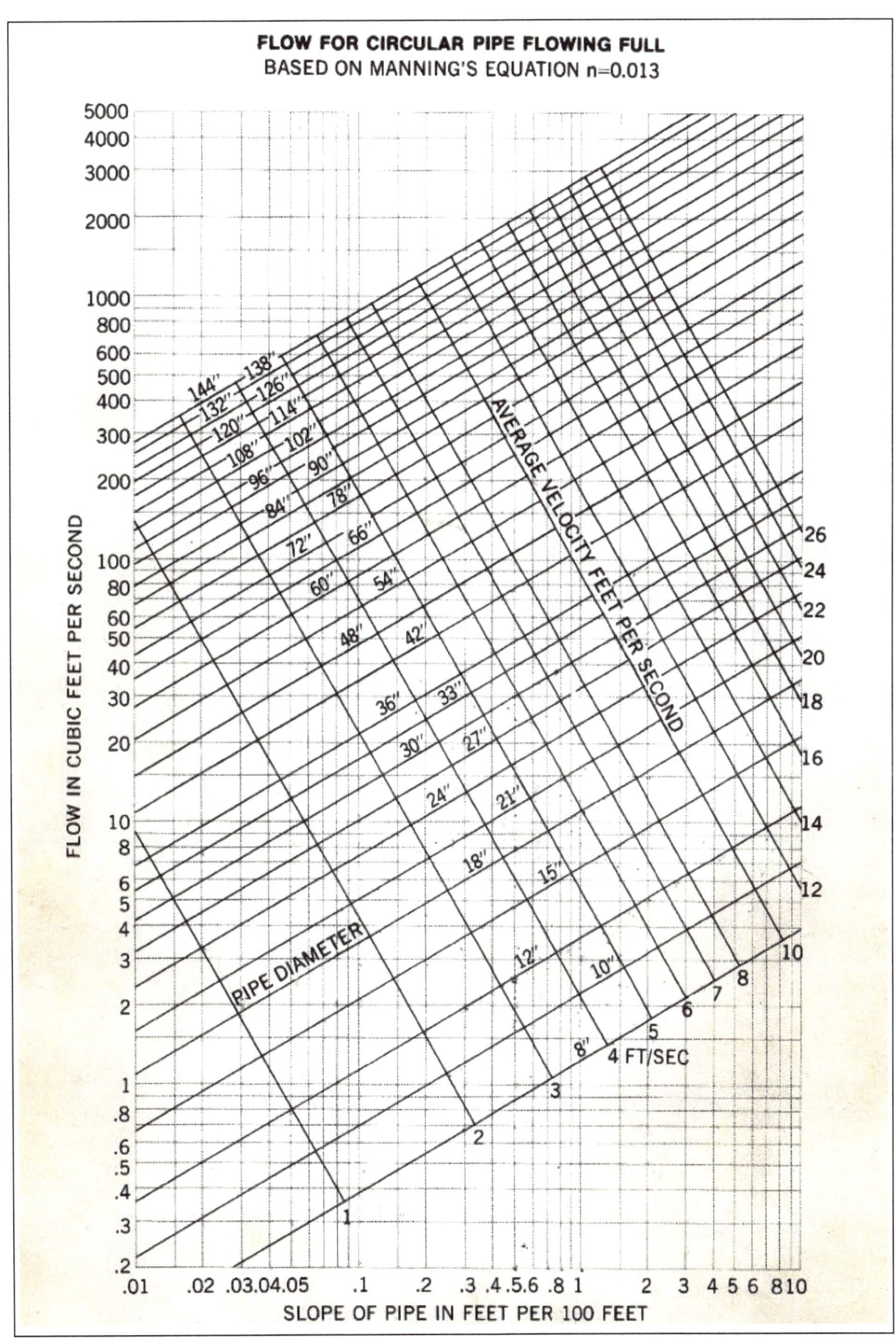

Today's computer programs solve the formulas, so Nomographs are no longer used, but civil engineers still study the variables – the amount of rain, what percentage of it runs off (as opposed to how much is absorbed into the ground), how rough the pipe is (the rougher it is, the smaller the capacity), what is the pipe size and slope (the capacity of flatter and smaller pipes is less than steeper and larger), and how much head can be provided.

The Project civil engineer studied these issues to design the storm sewers and ditches and the contractor installed them, based on the Plans. The civil engineer inspected the construction, they verified that the pipe and drainage structures met the specifications, and they checked the yield of aggregate supplied to backfill the trench.

Drainage structure with open grate to collect and transfer it to the storm sewer pipe:

12" storm sewer pipe being installed:

12" storm sewer system with aggregate backfill:

60" storm sewer shown only for reference. It was not installed on the Project:

Curb and Gutter: Type B-6.24 Concrete Curb and Gutter (24" wide gutter and 6" high barrier curb) was installed because the ROW limited construction in the dense urban area. It controlled drainage, improved safety, helped define the edge of pavement, and protected it from deterioration.

The civil engineer checked the yield by multiplying the C&G area times the length, converted it to cubic yards, and verified that enough concrete was delivered.

Barrier Curb and gutter (Type B-6.24 C&G) with 24" Gutter and 6" curb.
The elevations were on the Plans and were staked by the surveyor.

Testing Concrete: The civil engineer tested the concrete used for the Project C&G.

The Project had only one mix design, so it was simple, but many projects specify several different mixes, with different combinations of cement, aggregate, water, and chemical additives. Civil engineers, specializing in materials, working for the contractor, submit mix designs to show how the materials would be combined to provide the concrete that met the contract documents. The CM Team material coordinator would review the designs, work with the contractor to correct any deficiencies, and recommend approval to the owner.

Concrete testing required four different field tests -- for air content, slump, strength, and temperature. Each test required a different piece of equipment and an experienced civil engineer to perform them. They were completed by the QC Inspector as directed in the contract documents. In some specific, sensitive cases, such as bridge decks, civil engineers for the material supplier would also perform the tests. The tests were completed from a random selection of trucks, usually every 50-cubic yards.

Testing Concrete for Air Content:

The canister was loaded for the air test. It was filled at one-thirds with 25 plunges to properly consolidate.

The canister was loaded, smoothed, cleaned, and ready to cover with the meter.

The filled and pressurized canister showing the percentage of air. Between 5 and 8% was acceptable.

Testing Concrete for Slump:

The cone was loaded to test the slump. It was filled at one-thirds with 25 plunges to properly consolidate.

Slump began as soon as the cone was removed.

The slump was measured. Slump between 2" and 6" was acceptable.

Testing Concrete for Strength:

Cylinders were marked for identification.

The cylinders were filled at one-thirds with 25 plunges to properly consolidate.

Cylinders were stored in an ice chest against the cold to cure for 3, 7, or 14 days and broken at a laboratory by a civil engineer to determine their strength.

Testing Temperature of Concrete from Truck:

The temperature was checked – to show 60 degrees.

Pavement Construction: After the shoulders were removed, the pavement was widened 6-feet on each side with 14 ¼ inches of HMA. Then the 8-foot shoulder was added on each side with 10 ½ inches of HMA. Then 1 ½ inches of the remaining existing bituminous pavement surface was milled off. Finally, the entire, widened pavement, shoulders, and milled pavement were surfaced with 1 1/2 inches of HMA surface.

The civil engineer checked the density of each lift and the yield - the actual area times the plan thickness at a unit weight supplied by the mix design.

After removing the existing shoulders, the contractor constructed the widening and shoulder:

The civil engineer measured the widening and shoulder to verify proper width and to check the yield:

The milling machine removed the existing bituminous surface:

Milling consisted of a machine to mill and load into a dump truck and a sweeper to clean the remaining surface:

The civil engineer measured the milled depth for 1 ½ inch:

The final 1 ½ inch lift was placed as the surface:

The civil engineer used a nuclear density device to verify that the material was properly compacted. The maximum lift was 4-inches, each lift was compacted by rollers, and tested to verify the density met the requirements.

The civil engineer checked the HMA density many times on each lift:

After the final surface was placed, tested, and accepted, the CM Team laid out the proposed pavement striping.

Finally, the pavement was properly striped using thermoplastic pavement marking:

The finished pavement surface was then ready to accept the traffic. Right and left turn lanes at the proposed entrance reduced traffic conflicts with through traffic while providing a safe turning movement into the facility.

www.ingramcontent.com/pod-product-compliance
Lightning Source LLC
Chambersburg PA
CBHW040746200526
45159CB00023B/1755